The Making of the Grand Canyon

MARY JO NICKUM

Aquitaine Ltd
Phoenix, Arizona

The Making of the Grand Canyon

MARY JO NICKUM

Aquitaine Ltd
Phoenix, Arizona

Copyright © 2019 Mary Jo Nickum

Printed in the United States of America All Rights Reserved

No part of this book may be used or reproduced by any means, graphic, electronic, or mechanical, including photocopying, recording, taping or by any information storage retrieval system without the written permission of the publisher, except in the case of brief quotations embodied in articles and reviews.

Reviewers may quote passages for use in periodicals, newspapers, or broadcasts provided credit is given to *The Making of the Grand Canyon* by Mary Jo Nickum and Aquitaine, Ltd.

Aquitaine, LTD ISBN: 978-0-9980858-5-2
Library of Congress Cataloging Number LCCN: 2020939298

Printed in the United States of America First Edition

DEDICATION

This book is dedicated to young people everywhere who
Don't like to or find it difficult to read and
who, therefore, live on the fringes of a happy, healthy life.
Learn to enjoy reading and the world can be yours.

Believe in yourselves

Table of Contents

Part 1 How Rivers Carve Canyons 1
    Chapter 1 The Physical Setting 3
    Chapter 2 Geologic Formations 15

Part 2 The Human Factor 29
    Chapter 4 Native American People 31
    Chapter 5 The Explorers 33
    Chapter 6 The National Park 41

Part 3 The Grand Canyon Today and Its Future 45
    Chapter 7 Tourism and its Effect on the Canyon 47
    Chapter 8 Other Proposed Uses 51
    Chapter 9 The Future 62

Glossary 73
Side Bars 77
List of Illustrations 78
Illustration Credits 81
Sources 83
About the Author 87

Part 1

How Rivers Carve Canyons

The Grand Canyon is 277 miles long, up to eighteen miles wide and is over a mile (6,093 feet) deep. Nearly two billion years of Earth's geological history (Side Bar 1) have been exposed as the Colorado River and its tributaries cut their channels through layer after layer of rock, while, at the same time, the Colorado Plateau was uplifted. While some aspects about the history of incision of the canyon are debated by geologists, several recent studies support the hypothesis that the Colorado River established its course through the area about 5 to 6 million years ago. Since that time, the Colorado River has driven the down-cutting of the tributaries and retreat of the cliffs, at the same time deepening and widening the canyon. The Grand Canyon of the Colorado River is a world-renowned showplace of geology. Geologic studies in the park began with the work of Newberry in 1858, and continue today. The Grand Canyon's excellent display of layered rock is invaluable in unraveling the region's geologic history. Extensive carving of the plateaus allows for the detailed study of the Earth's movements. Processes of stream erosion and vulcanism are also easily seen and studied.

Side Bar 1

Major events in a brief geological time scale

Some of the more significant events in a brief geological time scale (mya = million years ago) are listed in the table below:

12 000 – 18 000 mya	Formation of the universe by the 'Big Bang'
4540 mya	Formation of the solar system including the Earth
4000 mya	Early formation of continents and tectonic plates
3700 mya	First primitive unicellular life forms
1500 mya	First evidence of advanced cell structures
542-520 mya	Proliferation and diversification of multicellular life forms
490 mya	First green plants and fungi emerge on land
420 mya	First land animals
330 mya	Large primitive trees appear
320 mya	First reptiles
250 mya	Up to 95% of life on Earth becomes extinct
230 mya	First dinosaurs appeared
150 mya	Gondwanan land mass starts to break up
150 mya	Monotremes, marsupials and placental mammals appear
65 mya	Mass extinctions, end of the dinosaurs
55 mya	Indian plate starts to collide with Asia, Himalayas start to form
5 mya	First human like life forms
0.06 mya	First record of aboriginal tribes in Victoria
0.05 mya	Construction of the Egyptian pyramids
0.0002 mya	Arrival of first Europeans in Australia

Chapter 1

The Physical Setting

The geology of the Grand Canyon area includes one of the most complete and studied sequences of rock on Earth. The nearly 40 major **sedimentary rock** layers exposed in the Grand Canyon and in the Grand Canyon National Park area range in age from about 200 million to nearly 2 billion years old (Side Bar 2). Most were deposited in warm, shallow seas and near ancient, long-gone sea shores in western North America. Both marine and terrestrial sediments are represented, including fossilized sand dunes from an extinct desert. There are at least fourteen known unconformities in the geologic record found in the Grand Canyon. An unconformity (Figure 1) represents time during which no sediments were preserved in a region. The local record for that time interval is missing and geologists must use other clues to discover that part of the geologic history of that area. The interval of geologic time not represented is called a *hiatus*.

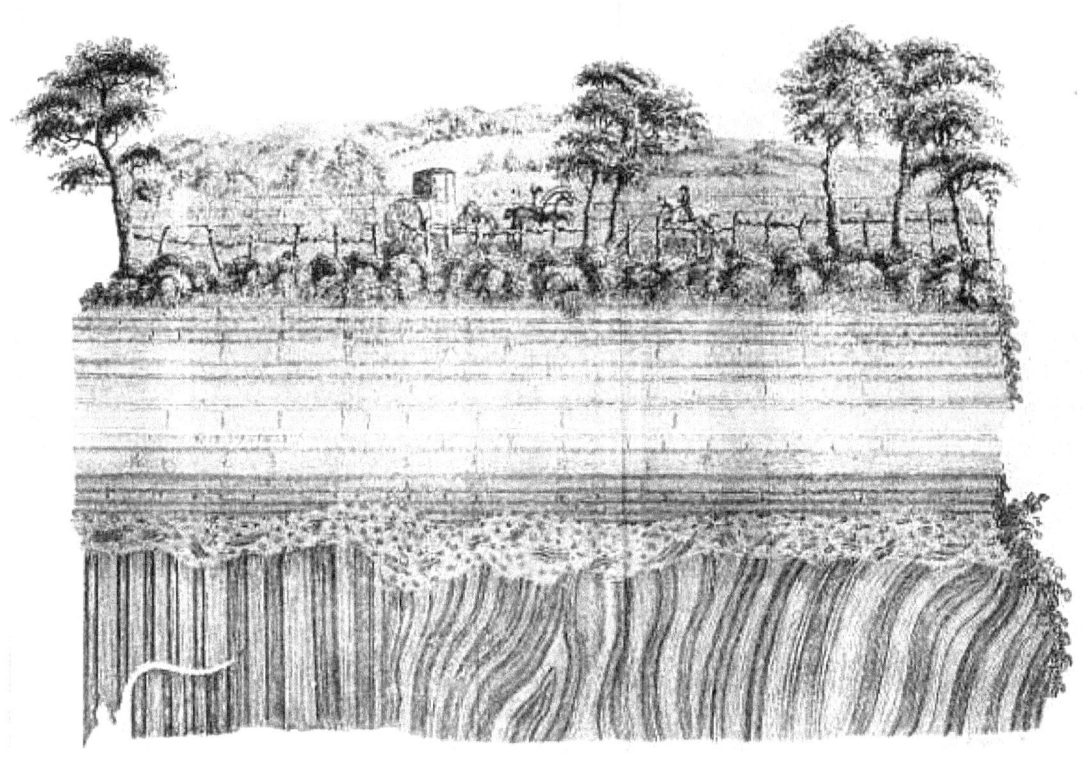

Fig 1. Unconformity

Uplift of the region started about 75 million years ago during the Laramide orogeny (Figure 2); a mountain-building event that is largely responsible for creating the Rocky Mountains to the east. In total, the Colorado Plateau was uplifted an estimated two miles. The adjacent Basin and Range Province to the west started to form about eighteen million years ago as the result of crustal stretching (Figure 3). Figure 4 provides a larger view of the Earth's crust. A drainage system that flowed through what is today the eastern Grand Canyon emptied into the now lower Basin and Range province. The opening of the Gulf of California, around 6 million years ago, enabled a large river to cut its way northeast from the gulf. The new river captured the older drainage to form the ancestral Colorado River (Figure 5), which in turn started to form the Grand Canyon. Rivers can flow down mountains, through valleys (depressions) or along plains, and can create canyons or gorges.

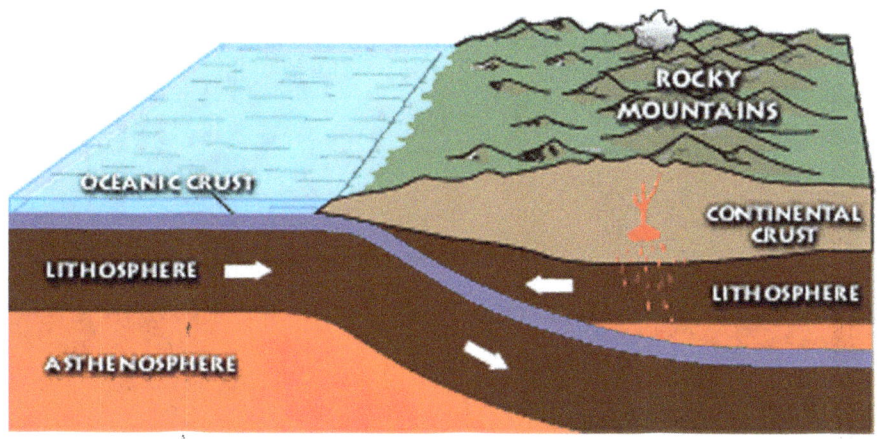

Fig 2. Shallow subduction -Laramide orogeny

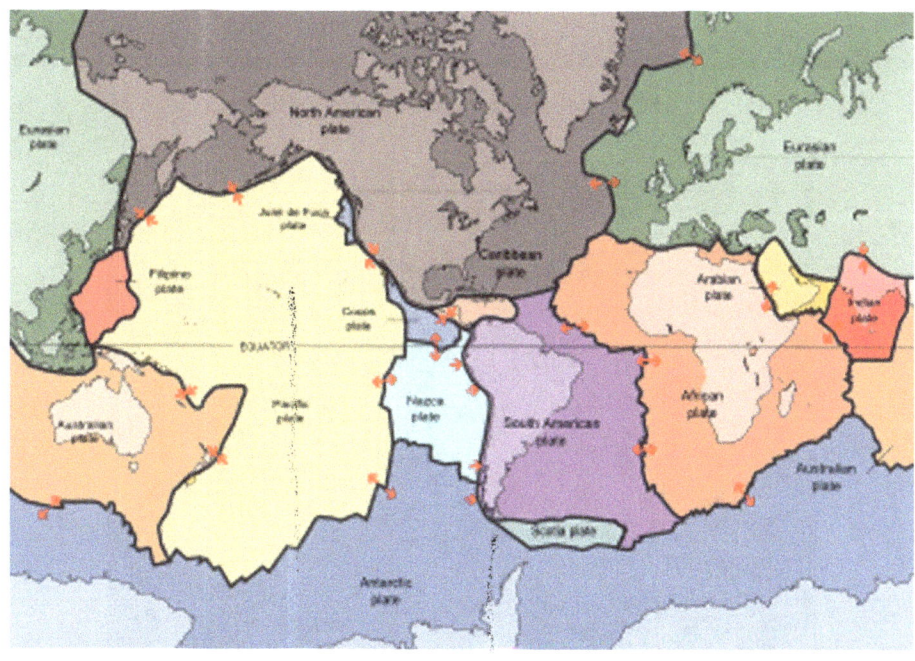

Fig 3. Plates of the crust of Earth

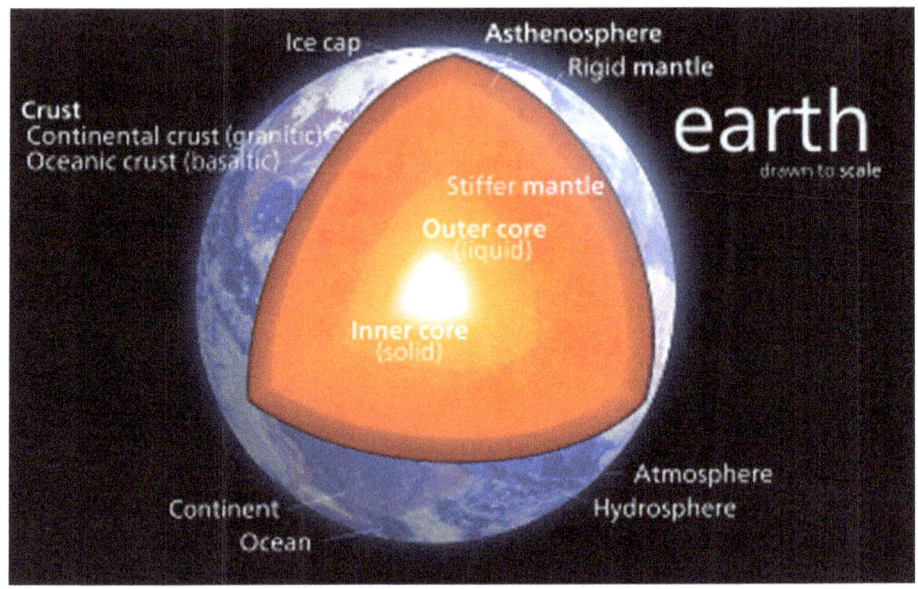

Fig 4. Internal Structure of the Earth

Fig 5. Colorado River

The Laramide orogeny affected all of western North America by helping to build the **American cordillera**. The Kaibab Uplift, Monument Upwarp, the Uinta Mountains, San Rafael Swell, and the Rocky Mountains were uplifted, at least in part, by the Laramide orogeny. This major mountain-building event started near the end of the Mesozoic, around 75 million years ago, and continued into the Eocene period of the Cenozoic (Side Bar 2). It was caused by subduction off the western coast of North America. Major faults that trend north–south and cross the canyon area were reactivated by this uplift. Many of these faults are Precambrian in age and are still active today. Streams draining the Rocky Mountains in early Miocene time terminated in landlocked basins in Utah, Arizona and Nevada but there is no evidence for a major river. Uplift of the Colorado Plateaus (Figure 6) forced rivers to cut down faster.

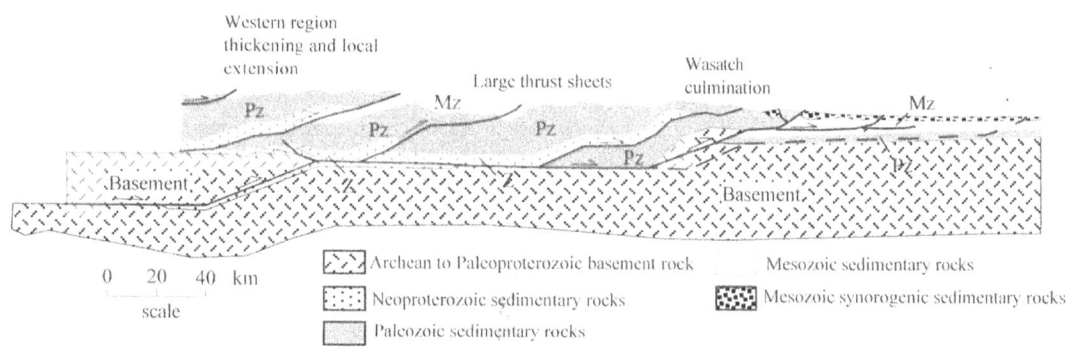

Fig 6. Colorado Plateau Uplift Cross-section (1 km = 0.6 miles)

Around 18 million years ago, tensional forces started to thin and drop the region to the west, creating the Basin and Range Province. Basins (**grabens**) dropped down and mountain ranges (**horsts**) rose up between old and new north–south-trending faults. However, for reasons poorly understood, the beds of the Colorado Plateaus remained mostly horizontal through both events even as they were uplifted about two miles in two pulses. The extreme western part of the canyon ends at one of the Basin and Range faults, the

Side Bar 2: Geologic Time Scale Explained

The Geologic time scale is a standard timeline used to describe the age of rocks and fossils, and the events that formed them.

Phanerozoic (542.0 mya to present)	Cenozoic (65.5 mya to present)	Quaternary (2.588 mya to present) • Holocene (11,700 yrs to present) • Pleistocene (2.588 mya to 11,700 yrs) Neogene (23.03 to 2.588 mya) • Pliocene (5.332 to 2.588 mya) • Miocene (23.03 to 5.332 mya) Paleogene (65.5 to 23.03 mya) • Oligocene (33.9 to 23.03 mya) • Eocene (55.8 to 33.9 mya) • Paleocene (65.5 to 55.8 mya)
	Mesozoic (251.0 to 65.5 mya)	Cretaceous (145.5 to 65.5 mya) • Upper (99.6 to 65.5 mya) • Lower (145.5 to 99.6 mya) Jurassic (199.6 to 145.5 mya) • Upper (161.2 to 145.5 mya) • Middle (175.6 to 161.2 mya) • Lower (199.6 to 175.6 mya) Triassic (251.0 to 199.6 mya) • Upper (228.7 to 199.6 mya) • Middle (245.9 to 228.7 mya) • Lower (251.0 to 245.9 mya)
		Permian (299.0 to 251.0 mya) • Lopingian (260.4 to 251.0 mya) • Guadalupian (270.6 to 260.4 mya) • Cisuralian (299.0 to 270.6 mya) Carboniferous (359.2 to 299.0 mya) • Pennsylvanian (318.1 to 299.0 mya) o Upper (307.2 to 299.0 mya)

		o Middle (311.7 to 307.2 mya)
		    o Lower (318.1 to 311.7 mya)
		• Mississippian (359.2 to 318.1 mya)
		    o Upper (328.3 to 318.1 mya)
		    o Middle (345.3 to 328.3 mya)
		    o Lower (359.2 to 345.3 mya)

	Paleozoic (542.0 to 251.0 mya)	Devonian (416.0 to 359.2 mya) • Upper (385.3 to 359.2 mya) • Middle (397.5 to 385.3 mya) • Lower (416.0 to 397.5 mya)
		Silurian (443.7 to 416.0 mya) • Pridoli (418.7 to 416.0 mya) • Ludlow (422.9 to 418.7 mya) • Wenlock (428.2 to 422.9 mya) • Llandovery (443.7 to 428.2 mya)
		Ordovician (488.3 to 443.7 mya) • Upper (460.9 to 443.7 mya) • Middle (471.8 to 460.9 mya) • Lower (488.3 to 471.8 mya)
		Cambrian (542.0 to 488.3 mya) • Furongian (499 to 488.3 mya) • Series 3 (510 to 499 mya) • Series 2 (521 to 510 mya) • Terreneuvian (542.0 to 521 mya)
Precambrian (4600 to 542.0 mya)	Proterozoic (2500 to 542.0 mya)	Neoproterozoic (1000 to 542.0 mya) Mesoproterozoic (1600 to 1000 mya) Paleoproterozoic (2500 to 1600 mya)
	Archean (4000 to 2500 mya)	Neoarchean (2800 to 2500 mya) Mesoarchean (3200 to 2800 mya) Paleoarchean (3600 to 3200 mya) Eoarchean (4000 to 3600 mya)
	Hadean (4600 to 4000 mya)	

Grand Wash, which also marks the boundary between the two provinces. Uplift from the Laramide orogeny and the creation of the Basin and Range province worked together to steepen the **gradient** of streams flowing west on the Colorado Plateau. These streams cut deep, eastward-growing, channels into the western edge of the Colorado Plateau and deposited their sediment in the widening Basin and Range region.

Wetter climates brought upon by ice ages starting 2 million years ago greatly increased excavation of the Grand Canyon, which was nearly as deep as it is now. Volcanic activity deposited lava over the area 1.8 million to 500,000 years ago (Side Bar 3). At least thirteen lava dams blocked the Colorado River, forming lakes that were up to 2,000 feet deep. The end of the last ice age and subsequent human activity has greatly reduced the ability of the Colorado River to excavate the canyon. Dams, in particular, have upset patterns of sediment transport and deposition. Controlled floods from Glen Canyon Dam upstream have been conducted to see if they have a restorative effect. Earthquakes and mass wasting erosive events still affect the region. Mass wasting, also known as slope movement or mass movement, is the **geomorphic process** by which soil, sand, **regolith**, and rock move downslope typically as a mass, largely under the force of gravity, but frequently affected by water and water content as in submarine environments and mudflows

Rifting started to create the Gulf of California (Figure 7) far to the south 6 to 10 million years ago. Around the same time, the western edge of the Colorado Plateau may have sagged slightly. Both events changed the direction of many streams toward the sagging region and the increased gradient caused them to down cut much faster. From 5.5 million to 5 million years ago, head-ward erosion to the north and east consolidated these streams into one major river and associated tributary channels. This river, the ancestral Lower Colorado River, started to fill the northern arm of the gulf, which extended nearly to the site of Hoover Dam, with estuary deposits.

Fig 7. Gulf of California

At the same time, streams flowed from highlands in central Arizona north and across what is today the western Grand Canyon, possibly feeding a larger river. The mechanism by which the ancestral Lower Colorado River captured this drainage and the drainage from much of the rest of the Colorado Plateau is not known. Possible explanations include head-ward erosion or a broken natural dam of a lake or river. Whatever the cause, the Lower Colorado probably captured the landlocked Upper Colorado somewhere west of the Kaibab Uplift. The much larger drainage area and yet steeper stream gradient helped to further accelerate down-cutting.

Ice ages during the Pleistocene brought a cooler and wetter **pluvial** climate to the region starting 2 to 3 million years ago. The added precipitation increased runoff and the erosive ability of streams, especially from spring melt water and flash floods in summer. With a greatly increased flow volume, the Colorado cut faster than ever before and started to quickly excavate the Grand Canyon 2 million years before present, almost reaching the modern depth by 1.2 million years ago.

The resulting Grand Canyon of the Colorado River trends roughly east to west for 278 miles between Lake Powell and Lake Mead (Figure 8). In that distance, the Colorado River drops 2,000 feet and has excavated an estimated 1,000 cubic miles of sediment to form the canyon (Figure 9). This part of the river bisects the 9,000-foot-high Kaibab Uplift and passes seven plateaus—the Kaibab, Kanab, and Shivwits plateaus bound the northern part of the canyon and the Coconino bounds the southern part. Each of these plateaus is bounded by north to south trending faults and monoclines created or reactivated during the Laramide orogeny. Streams flowing into the Colorado River have since exploited these faults to excavate their own tributary canyons, such as Bright Angel Canyon (Figure 10).

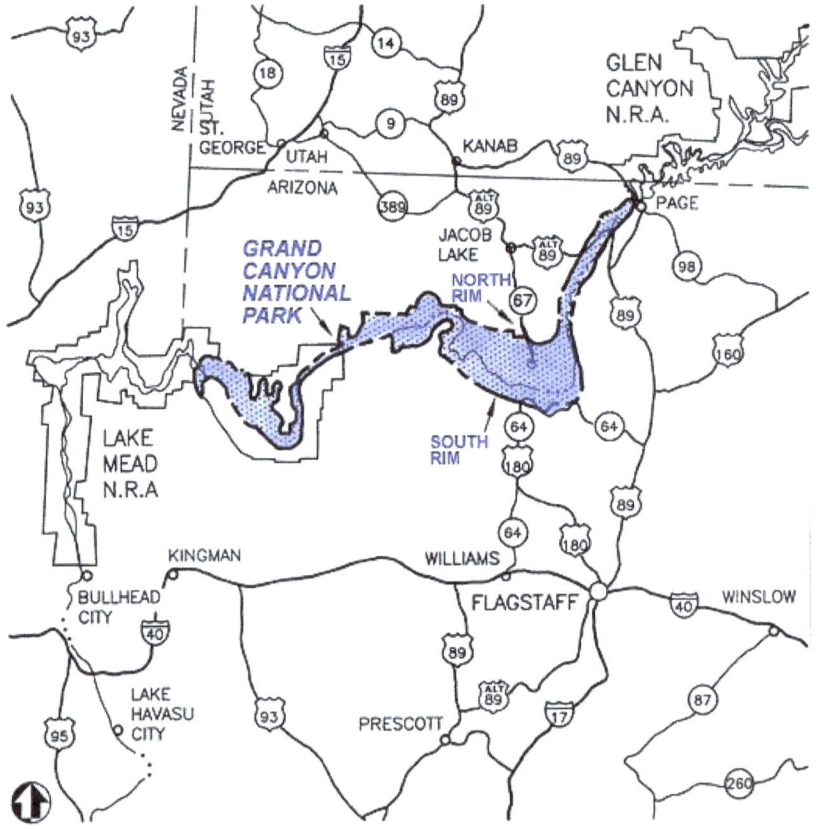

Fig 8. Grand Canyon Map

Fig 9. Sediments formed the Grand Canyon

Fig 10. Bright Angel Canyon

Chapter 2

Geologic Formations

The Colorado River has carved the Grand Canyon into four plateaus of the Colorado Plateau Province. The Province is a large area in the Southwest characterized by nearly-horizontal sedimentary rocks lifted 5,000 to 13,000 feet above sea level. The Plateau's arid climate produced many striking erosional forms, culminating in the Grand Canyon (Figure 11). The mile-high walls of the Canyon display a largely undisturbed cross section of the Earth's crust extending back some two billion years (Side Bar 4). Three "Granite Gorges" expose crystalline rocks formed during the early-to-middle **Proterozoic Era** (late **Precambrian**). Originally deposited as sediments and lava flows (Side Bar 5), these rocks were intensely metamorphosed about 1,750 million years ago. Magma rose into the rocks, cooling and crystallizing into granite, and welding the region to the North American continent.

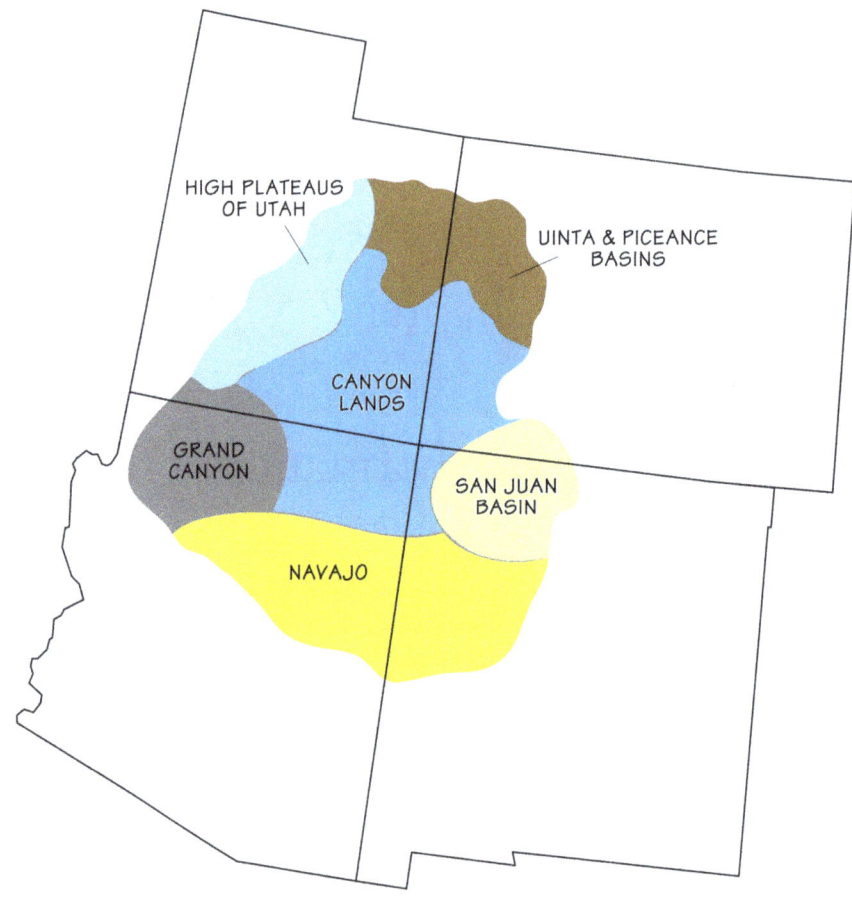

Fig 11. Sub-divisions of the Colorado Plateau

Side Bar 4

The rock record at Grand Canyon is summarized in the geologic cross section below.

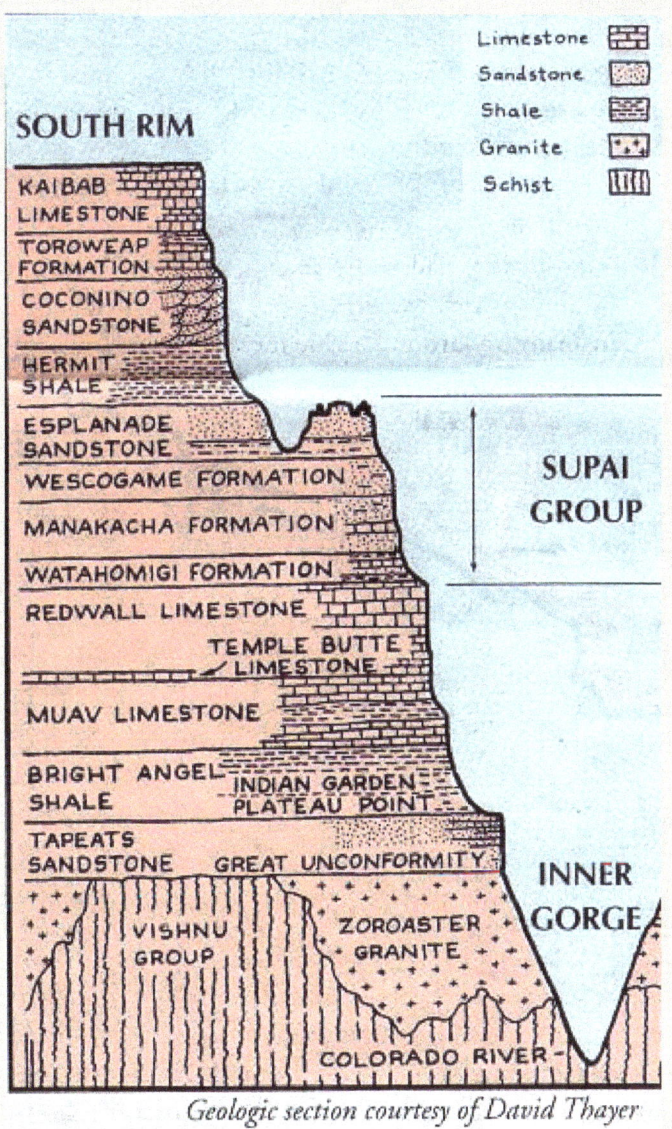

Geologic section courtesy of David Thayer

Side Bar 5: Time Periods by Name

The Mesozoic Era (See also Side Bar 2)

245-66.4 Million Years Ago

The Triassic Period
245-208 Million Years Ago

Throughout Triassic time, the Gulf Coastal area, including Florida, is land. In Late Triassic, the supercontinent Pangea begins to crack, break up and rift apart. Great grabens form, places where the lithosphere is foundering as this supercontinent rips apart. Late Triassic rocks mostly consist of sediments washing into those grabens from the bordering continents.

The Jurassic Period
208-144 Million Years Ago

In Early Jurassic time, the rocks are still terrestrial (deposited on land), representing continental sediments that are washing in the rift grabens that is formed as the Florida Platform separates from the African Plate. The Atlantic Ocean is opening. As North America also separates from South America, the Gulf of Mexico also begins to form, and sea water begins to flood the newly forming Gulf. Beginning in the Late Middle Jurassic, as the growing Gulf is intermittently shut off from oceanic waters, the waters of the Gulf evaporate, depositing over 3000' of evaporites (mainly salt). Eventually, by Late Jurassic, as the Gulf continues to open, circulation is no longer shut off and evaporites are no longer being deposited. Instead, carbonates (limestones and dolomites) are being put down as the first widespread ocean encroachment on land begins and continues into Cretaceous time.

The Cretaceous Period
144-66.4 Million Years Ago

In the Early Cretaceous, as seal level continues to rise, carbonate accumulation continues along the margin of the Florida platform. In the early part of the Middle Cretaceous, there is a rapid fall of sea level, soon followed by a rise, to eventually reach the one of the highest sea levels of all times. All along the East

Coast, including Florida, a carbonate platform very similar to the Great Barrier Reef develops (except immensely larger) and limestone deposition and reef formation continue through most of the Cretaceous. Climates are warm, and in the Gulf of Mexico, warm currents flow from the Atlantic into the Pacific ocean.

This carbonate deposition ends before the end of the Cretaceous and there is about a 10 million year gap in the depositional record till deposition resumes in the Middle Paleocene. So, unfortunately, the Florida record is silent on the great extinction at the end of the Cretaceous and its causes.

The Cenozoic Era

66.4 Million Years Ago *to now*

The Cenozoic Era is divided into two parts, the older and longer lasting Tertiary Period and the Quaternary Period in which we live now. These Periods are in turn subdivided into Epochs, based on the changing life patterns of the times.

The Tertiary Period
66.4-1.6 Million Years Ago

During the first 35 million years of the Tertiary, carbonate deposition is dominant on the Florida Platform. During the later Tertiary, several uplift periods in the Appalachians, and the associated erosion, provide clastic sediments that will reach Florida after the closing of the Suwanee Straight in the Oligocene. After the Oligocene, fluctuations in sea level, climate and ocean temperatures, associated with glaciations in different parts of the world and with altering plate tectonic patterns have been the major factors that controls sediment accumulation, distribution, and erosion, as well as the mode and tempo of the changing patterns of life on land and in the sea.

The Paleocene Epoch
66.4-57.8 Million Years Ago

Still isolated from influx of sediments (sands, silts and clays) derived from the lands to the North, bathed by warm northward-moving tropical waters, the

Florida Platform continues to accumulate carbonates, as it did in the Cretaceous. After the hiatus in the early Paleocene, sea level remains generally high and the platform commonly remains submerged. While the pattern of rock accumulation might be the same as during the Cretaceous, there is a major difference in life in the Paleocene seas as the result of the great extinction that took place at the end of the Cretaceous. The ammonites are gone. The great reef builders of the past (like the rudists) have either become extinct or, greatly reduced like the corals (although we do find evidence of patch reefs but not of massive reefs). Clams and snails continue to be important, and bryozoans and echinoderms flourish. It is also worth noting that elsewhere this is the time when the first grasses appear on land, although not in Florida, where there is no land. The Cedar Keys Limestone is deposited from the Late Paleocene into the Early Eocene.

The Eocene Epoch
57.8-36.6 Million Years Ago

In the Early Eocene North America and Europe are still joined. The Florida Platform is commonly submerged under a warm ocean and limestone deposition continues uninterrupted from the Paleocene as exemplified by the Cedar Keys Limestone.

In the Middle Eocene, the Oldsmar Limestone and overlying Avon Park Limestone are being put down in warm shallow seas where life still shows marked Tethyan (mediterranean) influences. This is the first time that massive coral reefs re-appear on the platform, although they sill are uncommon. Echinoderms continue to branch out and sand dollars appear, as do sea turtles. We also see the first appearance of archeocete whales, which still have large serrated teeth. Two genera are relatively common; Zygorhiza reaches up to 20 feet in length and the large Basilosaurus up to 70'. (More on fossil whales)

In Late Eocene, limestone continues to accumulate, only interrupted by occasional sea level fluctuations. The Ocala Limestone which succeeds the Avon Park, shows very little lithologic variation indicating that conditions are fairly uniform throughout that time. The sirenian Protosiren ancestor of the modern manatees and dugongs is found in the Ocala Ls.

> Some 40 Million Years ago, toward the end of the Eocene Epoch, the global circulation patterns that prevailed since Cretaceous time begin to change and uniformly warm ocean temperatures become a thing of the past, especially in the ever widening Atlantic. Two events work together to bring about this change. Firstly, the eastern Tethys seaway almost completely closes. This shuts off the warm currents that once upon a time practically circled the globe from East to West. Secondly, as North America separates from Europe, the northern part of the Atlantic Ocean opens. As it becomes connected to the Arctic Ocean, cold arctic ocean waters spill into the North Atlantic. Consequently, the abyssal (deepest part) of the ocean fills with cold water and cools significantly. This injection of cold water may be the main cause of the extinction of many marine forms that occurs at the end of the Eocene.

Beginning about 1,200 million years ago (late Proterozoic), 13,000 feet of sediment and lava were deposited in coastal and shallow marine environments. Mountain building about 725 million years ago lifted and tilted these rocks. Subsequent erosion removed these tilted layers from most areas leaving only the wedge-shaped remnants seen in the eastern Canyon (Figure 12).

Fig 12. Eastern Grand Canyon

The Ordovician and the Silurian, are missing from the Grand Canyon sequence. Geologists do not know if sediments were deposited in these periods and were later removed by erosion or if they were never deposited in the first place. Either way, this break in the geologic history of the area spans about 65 million years. A type of **unconformity** called a **disconformity** was formed. Disconformities show erosional features such as valleys, hills and cliffs that are later covered by younger sediments.

Geologists do know that deep channels were carved on the top of the Muav Limestone during this time. Streams were the likely cause, but marine scour may be to blame. Either way, these depressions were filled with freshwater limestone about 385 million years ago in the Middle Devonian in a formation that geologists call the Temple Butte Limestone. Marble Canyon in the eastern part of the park displays these filled purplish-colored channels well. Temple Butte Limestone is a cliff-former in the western part of the park where it is gray to cream-colored **dolomite**. Fossils of animals with backbones are found in this formation; bony plates from freshwater fish in the eastern part and numerous marine fish fossils in the western part. Temple Butte (Figure 13) is 100 to 450 feet thick; thinner near Grand Canyon Village and thicker in western Grand Canyon. An unconformity representing 40 to 50 million years of lost geologic history marks the top of this formation.

Fig 13. Temple Butte Limestone

The next formation in the Grand Canyon geologic column is the cliff-forming Redwall Limestone (Figure 14), which is 400 to 800 feet thick. Redwall is composed of thick-bedded, dark brown to bluish gray limestone and dolomite in which white chert nodules are mixed. It was laid down in a retreating shallow tropical sea near the equator during 40 million years of the early-to-middle Mississippian. Many fossilized marine organisms have been found in the Redwall. In late Mississippian time, the Grand Canyon region was slowly uplifted and the Redwall was partly eroded away. A **Karst topography** consisting of caves, sinkholes, and subterranean river channels resulted but were later filled with more limestone. The exposed surface of Redwall gets its characteristic color from rainwater dripping from the iron-rich redbeds of the Supai and Hermit shale that lie above.

Fig 14. Redwall Limestone

Surprise Canyon Formation (Figure 15) is a sedimentary layer of purplish-red shale that was laid down in discontinuous beds of sand and lime above the Redwall. It was created in very late Mississippian and possibly in very earliest Pennsylvanian time as the land subsided and tidal **estuaries** filled river valleys with sediment. This formation only exists in isolated lenses that are 50 to 400 feet thick. Surprise Canyon was unknown to science until 1973 and can be reached only by helicopter. Fossil logs, other plant material and marine shells are found in this formation. An unconformity marks the top of the Surprise Canyon Formation and in most places this unconformity has entirely removed the Surprise Canyon and exposed the underlying Redwall.

Fig 15. Suprise Canyon Formation

An unconformity of 15 to 20 million years separates the Supai Group (Figure16a, b) from the previously deposited Redwall Formation. Supai Group was deposited in late Mississippian, through the Pennsylvanian and into the early Permian time, some 320 million to 270 million years ago. Both marine and non-marine deposits of mud, silt, sand and calcareous sediments were laid down on a broad coastal plain similar to the Texas Gulf Coast of today. Around this time, the Ancestral Rocky Mountains rose in Colorado and New Mexico and streams brought eroded sediment from them to the Grand Canyon area.

Fig 16. Supai Group

Fig 17. Supai Group

Supai Group formations in the western part of the canyon contain limestone, indicative of a warm, shallow sea, while the eastern part was probably a muddy river delta. This formation consists of red siltstones and shale capped by tan-colored sandstone beds that together reach a thickness of 600 to 700 feet. Shale in the early Permian formations in this group were oxidized to a bright red color. Fossils of amphibian footprints, reptiles (Figure 17), and plentiful plant material (Figure 18) are found in the eastern part and increasing numbers of marine fossils are found in the western part.

Fig 18. Animal tracks

Fig 19. Fern fossil

Formations of the Supai Group are from oldest to youngest (an unconformity is present at the top of each): Watahomigi is a slope-forming gray limestone with some red **chert** bands, sandstone, and purple siltstone that is 100 to 300 feet thick. Manakacha is a cliff- and slope-forming pale red sandstone and red shale that averages 300 feet thick in Grand Canyon. Wescogame is a ledge- and slope-forming pale red sandstone and siltstone that is 100 to 200 feet thick. Esplanade is a ledge- and cliff-forming pale red sandstone and siltstone that is 200 to 800 feet thick. An unconformity marks the top of the Supai Group.

Part 2

The Human Factor

For thousands of years, the area has been continuously inhabited by Native Americans, who built settlements within the canyon and its many caves. The Pueblo people considered the Grand Canyon a holy site and made pilgrimages to it. The first European known to have viewed the Grand Canyon was García López de Cárdenas from Spain, who arrived in 1540.

Chapter 4

Native American People

The Ancestral Puebloans were a Native American culture centered on the present-day Four Corners area (Figure 19) of the United States. They were the first people known to live in the Grand Canyon area. The cultural group has often been referred to in archaeology as the Anasazi, although the term is not preferred by the modern Puebloan peoples. The word 'Anasazi' is Navajo for 'Ancient Ones' or 'Ancient Enemy'.

Fig 20. Four Corners of the Colorado Plateaus

Archaeologists still debate when this distinct culture emerged. The current consensus, based on terminology defined by the Pecos Classification (Side Bar 5), suggests their emergence was around 1200 BCE, during the Basketmaker II Era. Beginning with the earliest explorations and excavations, researchers have believed the Ancient Puebloans were ancestors of the modern Pueblo peoples.

In addition to the Ancestral Puebloans, a number of distinct cultures have inhabited the Grand Canyon area. The Cohonina lived to the west of the Grand Canyon, between 500 and 1200 CE. The Cohonina were ancestors of the Yuman, Havasupai, and Walapai peoples who inhabit the area today.

The Sinagua were a cultural group occupying an area to the southeast of the Grand Canyon, between the Little Colorado River and the Salt River, between approximately 500 and 1425 CE. The Sinagua may have been ancestors of several Hopi clans.

By the time of the arrival of Europeans in the 16th century, newer cultures had evolved. The Hualapai inhabit a 100-mile stretch along the pine-clad southern side of the Grand Canyon. The Havasupai have been living in the area near Cataract Canyon since the beginning of the 13th century, occupying an area the size of Delaware. The Southern Paiutes live in what is now southern Utah and northern Arizona. The Navajo, or Diné, live in a wide area stretching from the San Francisco Peaks eastwards toward the Four Corners. Archaeological and linguistic evidence suggests the Navajo descended from the Athabaskan people near Great Slave Lake, Canada, who migrated after the 11th century.

Chapter 5

The Explorers

Spanish Explorers

In September 1540, under orders from the conquistador Francisco Vázquez de Coronado to search for the fabled Seven Cities of Cibola, Captain García López de Cárdenas, along with Hopi guides and a small group of Spanish soldiers, traveled to the south rim of the Grand Canyon between Desert View and Moran Point. Pablo de Melgrossa, Juan Galeras, and a third soldier descended one third of the way into the canyon until they were forced to return because of lack of water. In their report, they noted some of the rocks in the canyon were "bigger than the great tower of Seville, Giralda". It is speculated their Hopi guides likely knew routes to the canyon floor, but may have been reluctant to lead the Spanish to the river. No Europeans visited the canyon again for more than two hundred years.

Fathers Francisco Atanasio Domínguez and Silvestre Vélez de Escalante were two Spanish priests who, with a group of Spanish soldiers, explored southern Utah and traveled along the north rim of the canyon in Glen and Marble Canyons in search of a route from Santa Fe to California in 1776. They eventually found a crossing, formerly known as the "Crossing of the Fathers", which, today, lies under Lake Powell (Side Bar 6). Also in

1776, Fray Francisco Garces, a Franciscan missionary, spent a week near Havasupai, unsuccessfully attempting to convert a band of Native Americans to Christianity. He described the canyon as 'profound'.

Side Bar 6: The "Crossing of the Fathers", now lies under Lake Powell

Canyon Country: This is a possible location where Native Americans guided the Spanish Fathers across the Colorado River. This bend is near the location and shows where Richard Sprang looked for possible crossings.

American Exploration

In the early 1800s, trappers and expeditions sent by the U.S. government began to explore and map the Southwest, including the canyon. Although first afforded Federal protection in 1893 as a Forest Reserve and later as a National Monument, the Grand Canyon did not achieve National Park status until 1919, three years after the creation of the National Park Service. Today Grand Canyon National Park encompasses more than 1 million acres of land and receives close to 5 million visitors each year.

James Ohio Pattie, along with a group of American trappers and mountain men, may have been the next Europeans to reach the canyon, in 1826.

Jacob Hamblin, a Mormon missionary, was sent by Brigham Young in the 1850s to locate suitable river crossing sites in the canyon. Building good relations with local Native Americans, the Hualapai Nation and white settlers, he found the Crossing of the Fathers and the locations of what would become Lees Ferry in 1858 and Pearce Ferry (later operated by, and named for, Harrison Pearce)—only the latter two sites were suitable for ferry operation. He also acted as an advisor to John Wesley Powell before his second expedition to the Grand Canyon, serving as a diplomat between Powell and the local native tribes to ensure the safety of his party.

In 1857, Edward Fitzgerald Beale (Side Bar 7) was superintendent of an expedition to survey a wagon road along the 35th parallel from Fort Defiance, Arizona to the Colorado River (Figure 20). He led a small party of men in search of water on the Coconino Plateau, near the canyon's south rim. On September 19, near present-day National Canyon, they came upon what May Humphreys Stacey described in his journal as "…a wonderful canyon four thousand feet deep. Everyone in the party admitted that they never seen anything to match or equal this astonishing natural curiosity."

Fig 21. Wagon road along the 35th parallel

Also in 1857, the U.S. War Department asked Lieutenant Joseph Ives to lead an expedition to assess the feasibility of an up-river navigation from the Gulf of California. Also in a stern wheeler steamboat *Explorer*, after two months and 350 miles of difficult navigation, his party reached Black Canyon two months after George Johnson. The *Explorer* struck a rock and was abandoned. Ives led his party east into the canyon—they may have been the first Europeans to travel the Diamond Creek drainage and traveled eastward along the south rim. In his "Colorado River of the West" report to the Senate in 1861, he stated "One or two trappers profess to have seen the canyon".

According to the *San Francisco Herald*, in a series of articles run in 1853, Captain Joseph R. Walker in January 1851 with his nephew James T. Walker and six men, traveled up the Colorado River to a point where it joined the Virgin River and continued east into Arizona, traveling along the Grand Canyon and making short exploratory side trips along the way. Walker is reported to have said he wanted to visit the Moqui Indians, as the Hopi

Side Bar 7: Edward Fitzgerald "Ned" Beale

Edward Fitzgerald Beale (1822-1893)

Edward Fitzgerald "Ned" Beale (February 4, 1822 – April 22, 1893) was a national figure in 19th century America. He was naval officer, military general, explorer, frontiersman, Indian affairs superintendent, California rancher, diplomat, and friend of Kit Carson, Buffalo Bill Cody and Ulysses S. Grant. He fought in the Mexican–American War, emerging as a hero of the Battle of San Pasqual in 1846. He achieved national fame in 1848 in carrying to the east the first gold samples from California, contributing to the gold rush.

In the late 1850s, Beale surveyed and built Beale's Wagon Road, which many settlers used to move to the West, and which became part of Route 66 and the route for the Transcontinental railroad. As California's first Superintendent of Indian Affairs, Beale helped charter a humanitarian policy toward Native Americans in the 1850s.

were then called by the whites. He had met these people briefly in previous years, thought them exceptionally interesting and wanted to become better acquainted. The *Herald* reporter then stated, "We believe that Capt. Joe Walker is the only white man in this country that has ever visited this strange people."

In 1858, John Strong Newberry became, probably, the first geologist to visit the Grand Canyon.

In 1869, Major John Wesley Powell, whose name has become synonymous with the Grand Canyon, led the first expedition down the canyon (Figure 21). Powell set out to explore the Colorado River and the Grand Canyon. Gathering nine men, four boats and food for ten months, he set out from Green River, Wyoming on May 24. Passing through or portaging around a series of dangerous rapids, the group passed down the Green River to its confluence with the Colorado River, near present-day Moab, Utah and completed the journey with many hardships through the Grand Canyon on August 13, 1869. In 1871, Powell first used the term 'Grand Canyon'; previously it had been called the 'Big Canyon'.

Fig 22. Rapids in the Powell expedition

In 1889, Frank M. Brown wanted to build a railroad along the Colorado River to carry coal. He, his chief engineer, Robert Brewster Stanton, and fourteen others started to explore the Grand Canyon in poorly designed cedar wood boats, with no life preservers. Brown drowned in an accident near Marble Canyon (Figure 22); Stanton made new boats and proceeded to explore the Colorado all the way to the Gulf of California.

Fig 23. Marble Canyon

Chapter 6

The National Park

The Grand Canyon became an official national monument in 1908 and a national park in 1919.

Grand Canyon is unmatched throughout the world in the incomparable vistas it offers to visitors on the rim. It is not the deepest canyon in the world, the Barranca del Cobre in northern Mexico and Hells Canyon in Idaho are deeper, just to name two, but the Grand Canyon is known throughout the world for its overwhelming size and its intricate and colorful landscape. Geologically, it is important because of the thick sequence of ancient rocks that are beautifully preserved and exposed in the walls of the canyon (Figure 23). These rock layers record much of the early geologic history of the North American continent. Finally, it is one of the most spectacular examples of erosion in the world.

Fig 24. Canyon Walls

The history of the Grand Canyon region is just as interesting. Grand Canyon was largely unknown until after the Civil War. In 1869, Major John Wesley Powell, a one-armed Civil War veteran with a thirst for science and adventure, made a pioneering journey through the Canyon on the Colorado River. He accomplished this with nine men in four small wooden boats, only six men completed the journey. His party was, as far as we know, the first ever to make such a trip.

In the late 19th Century there was interest in the region because of its promise of mineral resources, mainly copper and asbestos, as it turned out. The first pioneer settlements along the rim came in the 1880s. Early residents soon discovered tourism was destined to be more profitable than mining and, by the turn of the century, Grand Canyon was a well-known tourist destination. Many of the early tourist accommodations were not so different from the mining camps from which they developed, and most visitors made the grueling trip from nearby towns to the South Rim by stagecoach.

In 1901, the railroad was extended from Williams, Arizona to the South Rim (Figure 24), increasing dramatically the development of formal tourist facilities at the South Rim. By 1905 the El Tovar Hotel stood where it does today, a world class hotel on the edge of the canyon. The Fred Harvey Company, known throughout the west for hospitality and fine food, continued to develop facilities at Grand Canyon, including Phantom Ranch, built in the inner canyon in 1922. Today Grand Canyon National Park receives close to five million visitors each year—a far cry from the annual visitation of 44,173 which the park received in 1919.

Fig 25. Grand Canyon Railroad

Grand Canyon became a national park to give it the best protection we, as a nation, have to offer. The mission of the National Park Service, here and elsewhere, is to preserve the park and all of its features, including the processes that created them, and to provide for its enjoyment by park visitors in a way that will leave the canyon unspoiled for future generations. Now, more than ever, we recognize how complex and difficult a task that can be.

Part 3

The Grand Canyon Today and Its Future

Grand Canyon National Park is one of the world's premier natural attractions, drawing about five million visitors per year. Overall, 83 percent were from the United States: California (12.2 percent), Arizona (8.9 percent), Texas (4.8 percent), Florida (3.4 percent) and New York (3.2 percent) represented the top domestic visitors. Seventeen percent of visitors were from outside the United States; the most prominently represented nations were the United Kingdom (3.8 percent), Canada (3.5 percent), Japan (2.1 percent), Germany (1.9 percent) and The Netherlands (1.2 percent). The South Rim is open all year round, weather permitting. The North Rim is generally open mid-May to mid-October.

Chapter 7

Tourism and its Effect on the Canyon

Issues Facing Park Managers at the Grand Canyon

Environmental issues of concern to park managers at Grand Canyon are as diverse as the park itself and include air quality, fire management, the impact of increased visitation and endangered species, to name just a few. We tend to think of national parks as islands in time and space but that's a dangerous illusion; more and more the forces that affect the integrity of the park ecosystem come from outside the park and are beyond the direct control of park managers.

The issue of air quality at the Canyon is a prime example. Many summer visitors to Grand Canyon find the view from the rim obscured by regional haze carried in from urban and industrial areas to the south and west, from far outside the park. Even in winter, when prevailing winds tend to carry cleaner air from the north, emissions of sulfur dioxide from local sources can drastically impair visibility at Grand Canyon.

Water is another major issue: most of the water that finds its way into the Canyon comes from outside the park and the flow of the Colorado River through Grand Canyon is directly controlled by the inlet of water from Glen

Canyon Dam (Figure 25), just 15 miles upstream from the park (Figure 26). In the past few years, a great deal of research has been directed at the effect of Glen Canyon Dam on the Colorado River in Grand Canyon. For years, we've been aware of the more obvious effects of the dam: colder water, carefully controlled flows, the absence of distinct seasonal fluctuations and greatly decreased sediment load. Only now are we beginning to understand the long term effects these changes have had on the system, as a whole.

Fig 26. Glen Canyon Dam

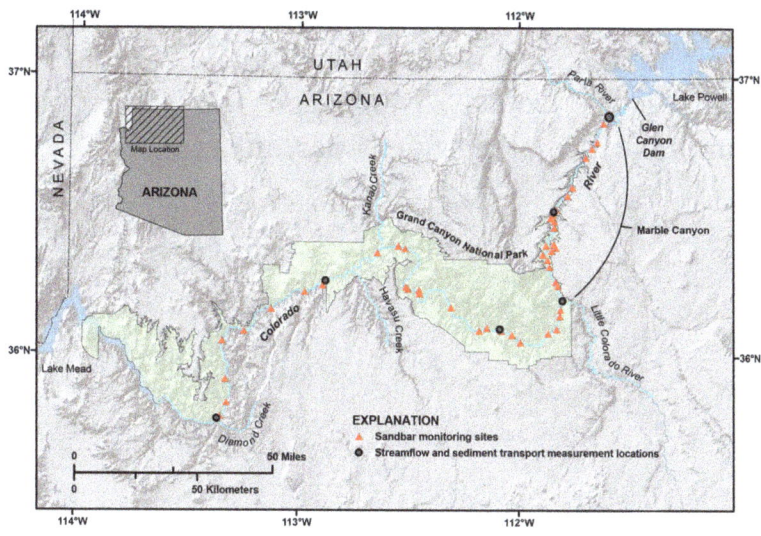

Fig 27. Glen Canyon and Grand Canyon Map

One of the larger projects using Colorado River water outside the Grand Canyon is the **Central Arizona Project** (CAP). This project is Arizona's single largest resource for renewable water supplies. The **Central Arizona Project** is designed to bring about 1.5 million acre-feet of water from the Colorado River to Central and Southern Arizona every year. More than 5 million people, or more than 80 percent of the state's population, live in Maricopa, Pima and Pinal counties, where CAP water is delivered. The **Central Arizona Project** carries water from Lake Havasu near Parker to the southern boundary of the San Xavier Indian Reservation southwest of Tucson. It is a 336-mile long system of aqueducts, tunnels, pumping plants and pipelines and is the largest single resource of renewable water supplies in Arizona.

In the early 20th century, Central Arizona Project was a shared dream of Arizonans; a vision of water security and stability for future generations to enjoy their quality of life in a desert. Now that the 336-mile long water delivery system is a reality, the leadership of CAP is responsible for protecting and preserving what past generations were able to fund and build.

Such issues are of interest to all of us, and there are no easy answers; but the task of preservation and protection within park boundaries is not nearly as simple as it must have looked 75 years ago. The issues that interest park managers are, in many cases, the same issues, affecting us all and ones we, as a nation, must address.

Chapter 8

Other Proposed Uses

Uranium Mining

In the 1980s, the US Bureau of Land Management (BLM) approved plans of operations for numerous uranium mines (Figure 27a, b) on the outskirts of Grand Canyon National Park. When uranium prices plummeted in the early 1990s, the operators placed the mines in a regulatory purgatory, known as "non-operation", where the mines sat idle for the next two decades. When uranium prices surged again, federal agencies allowed operations to resume without new environmental reviews.

Fig 27a. Grand Canyon Uranium Mine

Fig 27b. Uranium Mining-Havasupai Tribe

Working in a coalition with the Havasupai Tribe and other conservation groups, the Grand Canyon Trust sued the U.S. Forest Service (USFS) over its decision to allow the Canyon uranium mine to open without adequate tribal consultation or updating a 1986 federal environmental review.

Few people realize how close they are to a uranium mine when they visit Grand Canyon National Park. Located six miles south of Grand Canyon Village, Canyon Mine also sits within an area of religious and cultural importance to tribes, such as the Havasupai. The mine was located also above groundwater that supplied some of the most treasured seeps and springs of the Canyon, including Havasu Springs and Havasu Creek. After a 20-year hiatus, Canyon Mine planned to resume operations, but tribes and conservation groups were doing everything they could to stop it. Set to resume operations in 2018. The Canyon Uranium Mine owner, Energy Fuels, planned to truck uranium ore through Flagstaff and dozens of small communities enroute to White Mesa Uranium Mill in southeastern Utah (Figure 28).

Fig 28. White Mesa Uranium Mill

Grazing

On the Colorado Plateau and in the Grand Canyon, mammalian grazers affect ecosystems by altering vegetation composition, water quality and soil quality. The Grand Canyon is home to native and domesticated grazers, including the native mule deer (*Odocoileus hemionus*) and the desert bighorn sheep (*Ovis canadensis*), as well as cattle, sheep, goats, horses and bison (*Bison bison*). Today, livestock grazing in the Grand Canyon itself has ceased, but its legacy remains.

In addition, much of the surrounding area is still maintained as rangelands. The Grand Canyon ecosystem is held and maintained by a diverse set of stakeholders ranging from state and federal agencies to Native American communities to private landowners and conservation organizations. These stakeholders often disagree about how to manage native and domesticated grazers.

In general, livestock grazing is often linked to environmental degradation, water shortages, and forage scarcity. Livestock grazing also presents a number of threats to wildlife. Livestock grazing can increase competition with wildlife for food and water. Moreover, livestock grazing can decrease nutrient availability in the soil. In the Grand Canyon, grazing has been shown to reduce soil crusts.

Competition between wild herbivores and livestock is likely context-dependent. For example, in one grassland system in Kenya, wild herbivores and cattle compete for food during the dry season when resources are scarce, but facilitate each other's diet quality during the wet season when resources are high; however, the effects of livestock need not always be negative. In some cases, livestock may provide unexpected benefits to an ecosystem by promoting seed dispersal and increasing plant diversity.

Even though grazing may have both positive and negative effects on ecosystem health, the legacy of livestock grazing around the Grand Canyon is largely negative. This is likely related to widespread overgrazing at the end of the nineteenth century and the beginning of the twentieth.

The history of grazing began with the Spaniards, who brought livestock as early as the 1500s, but livestock grazing was not important in this area until the late 1800s. The arrival of railroads to the Colorado plateau also meant the arrival of hundreds of thousands of sheep and cattle. Although estimates of the exact number of sheep and cattle in and around the Grand Canyon at the end of the 1800s vary, it is largely agreed the stocking rates were much too high. For example, in 1881 alone, 20,000 head of cattle and 22,000 more arrived to the Colorado Plateau via railroad.

This overstocking likely had negative effects on vegetation, soil, and fire patterns in the region. In the adjacent Little Colorado Basin, also subject to extensive overgrazing, range quality deteriorated throughout the late 1800s. Overstocking led to severe erosion (Figure 29) and halted the regeneration of Ponderosa pine forests on the adjacent Kaibab Plateau. In 1906, President Theodore Roosevelt established the Grand Canyon Game Preserve. This designation as a game preserve restricted grazing in the Grand Canyon, although it was still legal. Grazing continued in the Grand Canyon until 1919, when the area was designated as a national park; however, the effects of grazing were still seen years after the Grand Canyon became a national park.

Fig 29. Effects of grazing and browsing on trees

Further grazing regulation in the Grand Canyon was established with the passage of the Taylor Grazing Act of 1934. This act increased the regulation of grazing on federal lands. In short, the days of overgrazing livestock in the Grand Canyon were largely over by the 1930s. In addition, surrounding federal lands also became more regulated. Today, the area surrounding the Grand Canyon is a patchwork of different agencies and stakeholders (Figure 30).

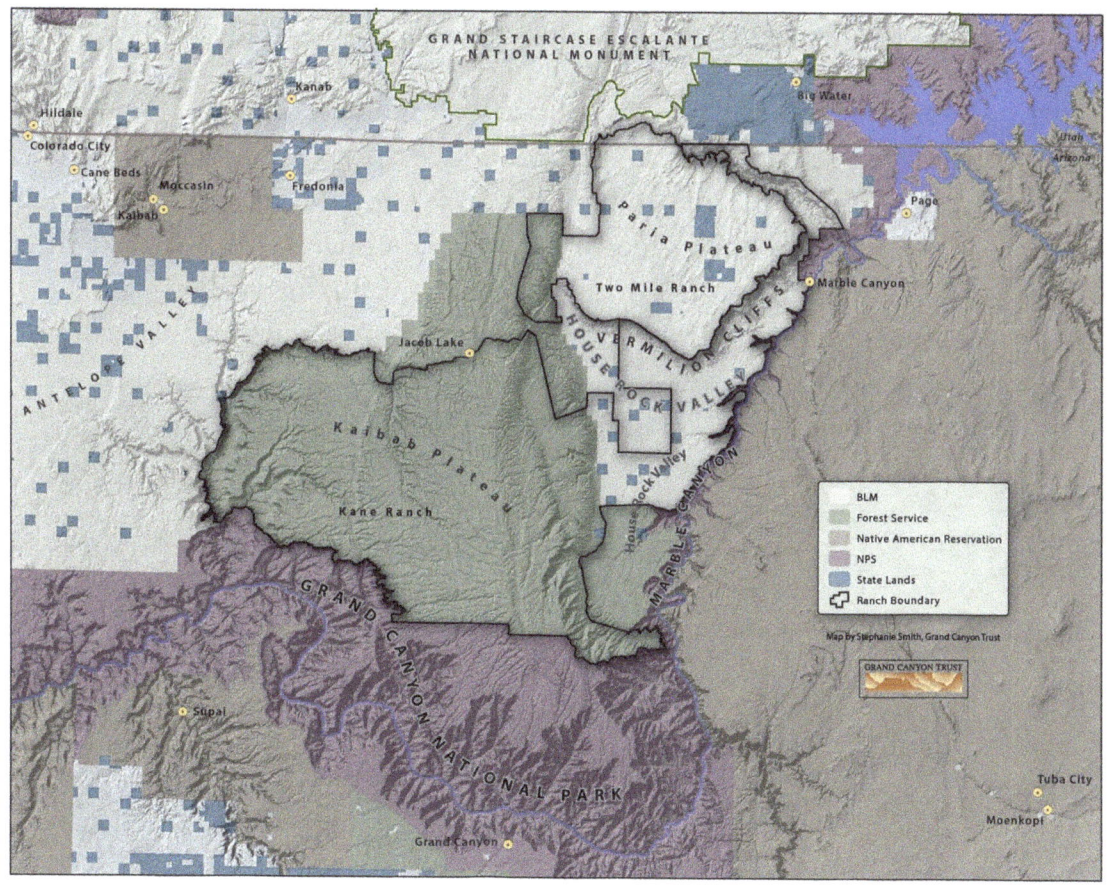

Fig 30. Canyon National Park is a patchwork of different management types

Grazing is still allowed in some areas around the Grand Canyon, including on some federal and state properties, private areas, and on Native American reservations; yet, management varies. For example, management of grazing on reservations is largely the jurisdiction of the Bureau of Indian Affairs (BIA).

Hunting

At the head of the Grand Canyon, near Page, Arizona, Lees Ferry is a fly-fishing dream. Think about reeling in rainbow, brown and cutthroat trout from a scenic Colorado River spot, while Arizona's diverse hunting options include larger game such as elk, as well as smaller game and waterfowl.

Sporting of a different sort involves the National Park Service (NPS) working with cooperating agencies and partners to reduce the size of the bison herd on the North Rim (Figure 31) of Grand Canyon National Park (Park). The herd is approximately 400 to 600 animals and will be reduced over the next three to five years to fewer than 200 by lethal culling, utilizing skilled volunteers and non-lethal capture and transfer to cooperating agencies and tribes.

Fig 31. Bison on the North Rim

Here is why... Given the current distribution, abundance, density, and the expected growth of this herd, the NPS is concerned about increased impacts on park resources, such as water, vegetation, soils, archaeological sites, and values, including visitor experience and wilderness character. Reducing the herd size will protect Park resources and values. The NPS biologists estimate the herd has grown from approximately 100 bison, brought to the House Rock Wildlife Area in the early 1900s, to between 400 to 600 bison. Though the bison roam the Kaibab Plateau, they spend most of their time on the North Rim of the park. Biologists predict the herd could grow to nearly 800 in the next three years and be as large as 1200 to 1500 animals within ten years, without further management actions to control the size of the herd.

Recreational Use Summarized

Over the years, human activities have impacted the natural resources of Grand Canyon National Park in many ways. Humans have introduced non-native plant and animal species into the park, which outcompete native flora and fauna for space, food and water. Air pollution has routinely drifted into the Canyon from metropolitan areas and nearby coal-fired power plants, affecting visibility from scenic vistas (Figure 32). Water in some streams has been tainted with fecal coliform from trespass cattle and from human waste. The construction of Glen Canyon Dam in 1963 irreversibly altered the riparian and aquatic ecosystems within the park. The natural quiet of Grand Canyon has been disturbed by rumbling aircraft noise, and forest landscapes have been altered by decades of wildland fire suppression.

Fig 32. Grand Canyon scenic vista

Today, many laws have been passed and programs put in place to protect and restore the natural wonders of the Grand Canyon "to leave them unimpaired for the enjoyment of future generations", Park scientists use integrated pest management techniques to eradicate and relocate non-native and pest species. A coal-fired power plant in Page (Figure 33) has installed scrubbers in smoke stacks to reduce air pollution. Fences have been erected along the park boundaries to keep out trespass cattle, and hikers and river runners are being educated on the proper methods of human waste disposal. Stakeholders from federal and state agencies, Native American tribes, and environmental and recreational organizations have partnered to create the Glen Canyon Dam Adaptive Management Program to recommend modifications to dam operations to benefit natural and cultural resources in Grand Canyon National Park and Glen Canyon National Recreation Area. Special no flight zones have been created to preserve natural quiet in remote areas of the park and prescribed burning and forest thinning are natural resource management tools used to restore forest landscapes and reduce wildfire hazards.

Fig 33. Coal-fired Power Plant in Page, AZ

Chapter 9

The Future

People's behavior throughout the West, particularly the Southwestern United States, was conditioned and circumscribed by the perennial shortage of water. The expected, but variable, supplies of surface water were quickly appropriated. Electricity and electric pumps enabled access to previously unavailable groundwater sources, while the favorable climate resulted in an increase in agriculture and urbanization. As a consequence, nearly all of the water supplied to this rapidly growing area was pumped from underground basins. This has caused a steady decline in regional water tables, which, in turn, has affected local economies. Many acres that formerly supported agriculture have been abandoned, converted to housing developments, or switched to an alternate water source, such as the Central Arizona Project (CAP; Side Bar 8), which became available in the late 1980s. The water situation, however, especially in heavily populated areas, has had little effect on people's water consumption, except for farmers. As the cost of water increases, the farmer's income decreases. Eventually, the farmer is forced to stop farming and, either abandons or sells the land. The profit margin for the urban home owner is much higher. Consequently, Arizona has many human-made lakes, golf courses, and green lawns, and residents continue to demand more; conversion of water previously used for agriculture, however, has the potential to sustain

Side Bar 8: Recommendations/Suggestions for water in the Grand Canyon National Park:

Presented below are some best practice recommendations and suggestions identified as part of a qualitative evaluation.

No or Minimal Cost

- Assess the need of the tertiary treatment of wastewater: it is rarely done in comparable systems.

- Discuss alternatives regarding the chlorination of drinking water: using helicopters to transport chlorine gas is energy consumptive.

- Coordinate raw water production/distribution with treatment process capacity.

- Assess costs of maintaining existing facilities versus upgrading over the expected life of the system.

- Review system plans, specifications, and records with plant operators, maintenance staff, and engineers before considering upgrades/improvements.

- Evaluate costs for different available water sources.

- Secure operations and maintenance guides and training for park staff when new systems/components are installed.

- Allow the bar screen rake to rise only enough to expose more clean bar screen, letting excess water drip off before deposition in a dumpster, thereby not hauling excess water weight to the landfill.

Low to Moderate Cost

- Implement a water conservation and education program.

- Evaluate pumps, blowers, and motors for upgrade to high-efficiency or VFD.

- Directly utilize heat in blower/pump room to keep biological activities warmer and thus more reactive or to heat nearby buildings in the winter.

- Develop a cost analysis and implement capital improvement planning.

- Reduce leakage through pressure management.

- Adopt water-efficient ordinances and codes.

- Conduct an energy audit of all pumps and blowers and their total energy consumption.

- Retrofit facilities with energy efficient lighting, using high-efficiency ballasts and bulbs.

- Perform a loss/leakage survey for both reclaimed water and influent wastewater.

- Create financial or other incentives for water users to conserve instead of increasing production/treatment capacity.

- Utilize off-peak power usage strategies.

- Adequately ventilate or sunshield all electrical and mechanical equipment in warm weather.

- Take measurements and evaluate the data prior to making future improvement plans.

Moderate to High Cost

- Reduce friction/energy losses in pumps, fans, pipes, valves, and production wells.

- Replace the entire TCP with a larger-diameter lower-friction-loss steel pipe that can handle washouts, increased future demands, and water pressure transients during valve closures or flow adjustments.

> - Reduce pressures within the new TCP with an electric turbine to generate supplemental electricity power.
>
> - Use a MIOX system at Roaring Springs: the system runs on salt, thus mules can safely bring in the raw materials.
>
> - Install a hydro-electric turbine on Bright Angel Creek, near Phantom Ranch, to generate power for Phantom Ranch and to run a UV-dechlorination unit, discharging surplus water to Bright Angel Creek, which Roaring Springs normally feeds.
>
> - Change coarse-bubble diffusers to fine-bubble diffusers to increase aeration efficiency.
>
> - Utilize renewable energy wherever appropriate throughout the water and wastewater systems.

the growth of municipalities and industry into the future (Side Bar 9).

The combined surface and ground water supplies in the Colorado River Basin are generally adequate for current needs; however, growing demands and uses of water in this basin could soon result in a widespread water shortage. Local shortages already exist. Barring conversion of saline water, additional importation of outside water, advancements in rainmaking, and rigorous conservation measures, residents must rely on the variable surface and diminishing groundwater supplies.

In response, the initial direction of the research in the Colorado River Basin (Figure 34) focused on investigating the potentials for increasing water yields from the region's forests, woodlands, and shrub lands through vegetative manipulations. Numerous watersheds were instrumented with climatic and hydrologic measuring devices by the US Forest Service and its cooperators in the late 1950s and throughout the 1960s to study the effects of vegetative clearings, thinnings, and conversion of vegetation on water yields under controlled, experimental conditions.

Side Bar 9: Recommendations/Suggestions for water in the Grand Canyon National Park:

Presented below are some best practice recommendations and suggestions identified as part of a qualitative evaluation.

No or Minimal Cost

- Assess the need of the tertiary treatment of wastewater: it is rarely done in comparable systems.

- Discuss alternatives regarding the chlorination of drinking water: using helicopters to transport chlorine gas is energy consumptive.

- Coordinate raw water production/distribution with treatment process capacity.

- Assess costs of maintaining existing facilities versus upgrading over the expected life of the system.

- Review system plans, specifications, and records with plant operators, maintenance staff, and engineers before considering upgrades/improvements.

- Evaluate costs for different available water sources.

- Secure operations and maintenance guides and training for park staff when new systems/components are installed.

- Allow the bar screen rake to rise only enough to expose more clean bar screen, letting excess water drip off before deposition in a dumpster, thereby not hauling excess water weight to the landfill.

Low to Moderate Cost

- Implement a water conservation and education program.

- Evaluate pumps, blowers, and motors for upgrade to high-efficiency or VFD.

- Directly utilize heat in blower/pump room to keep biological activities warmer and thus more reactive or to heat nearby buildings in the winter.

- Develop a cost analysis and implement capital improvement planning.

- Reduce leakage through pressure management.

- Adopt water-efficient ordinances and codes.

- Conduct an energy audit of all pumps and blowers and their total energy consumption.

- Retrofit facilities with energy efficient lighting, using high-efficiency ballasts and bulbs.

- Perform a loss/leakage survey for both reclaimed water and influent wastewater.

- Create financial or other incentives for water users to conserve instead of increasing production/treatment capacity.

- Utilize off-peak power usage strategies.

- Adequately ventilate or sunshield all electrical and mechanical equipment in warm weather.

- Take measurements and evaluate the data prior to making future improvement plans.

Moderate to High Cost

- Reduce friction/energy losses in pumps, fans, pipes, valves, and production wells.

- Replace the entire TCP with a larger-diameter lower-friction-loss steel pipe that can handle washouts, increased future demands, and water pressure transients during valve closures or flow adjustments.

- Reduce pressures within the new TCP with an electric turbine to generate supplemental electricity power.

- Use a MIOX system at Roaring Springs: the system runs on salt, thus mules can safely bring in the raw materials.

- Install a hydro-electric turbine on Bright Angel Creek, near Phantom Ranch, to generate power for Phantom Ranch and to run a UV-dechlorination unit, discharging surplus water to Bright Angel Creek, which Roaring Springs normally feeds.

- Change coarse-bubble diffusers to fine-bubble diffusers to increase aeration efficiency.

- Utilize renewable energy wherever appropriate throughout the water and wastewater systems.

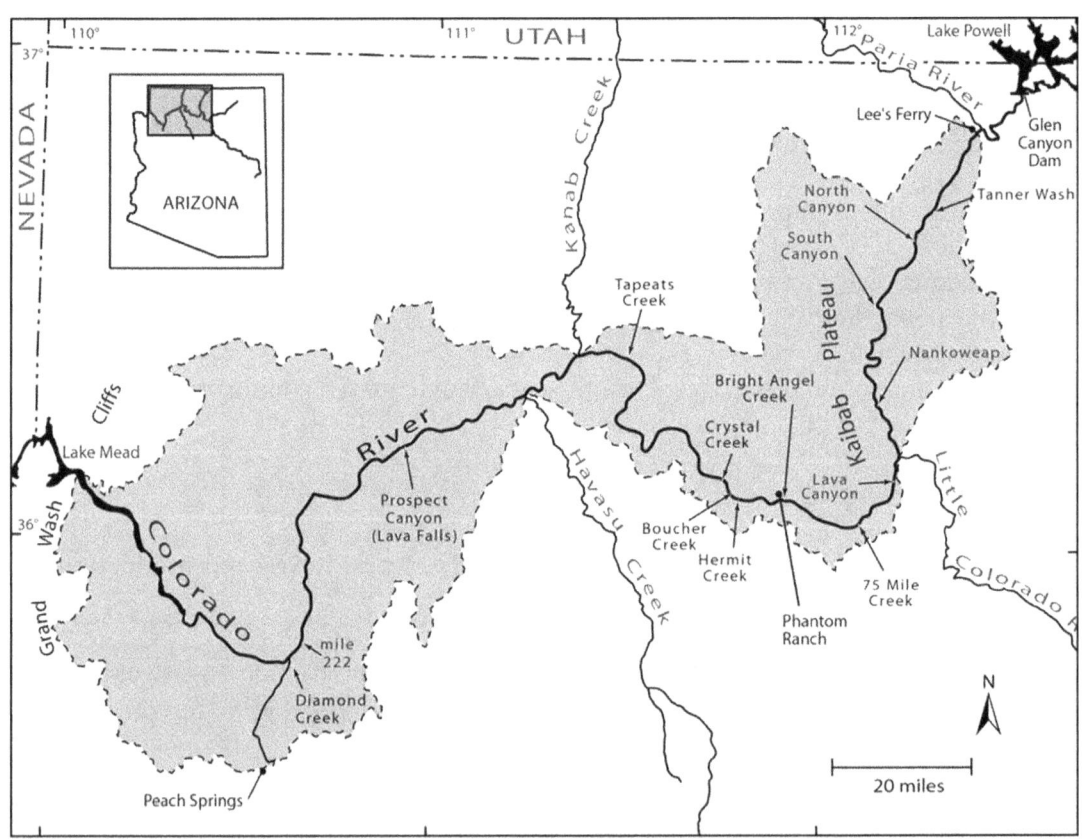

Fig 34. The Colorado River Basin in Grand Canyon, Arizona

Theoretically, the surface water supply in the Colorado River Basin could be increased by as much as 1/3 if vegetation and snow on 16 percent of the basin were manipulated solely to increase water yield. Other forest resources, economics, and social and environmental concerns, however, would greatly reduce the treatment area and effectiveness of the increasing water yield.

Water-yield increases are greatest where large reductions can be made in water transpired by plants (Side Bar 10) and evaporated from snow. **Clearcutting** (Figure 35) and conversion of vegetation usually increase water yield greatly. These practices can be appropriate in several vegetation types, such as chaparral and mountain brush, where the commercial value of the vegetation is low. Where clear-cuts and type conversions are unacceptable management practices, the potential for increasing the water yield is less, although it can still be substantial.

Fig 35. Clear cutting

Researchers report water yield in the Upper Colorado River Basin could be increased by 152,340 acres per year, or 3.5 percent, by treating up to 22 percent of each vegetation type, except aspen (*Populus tremuloides*) where 40 percent would be treated. About half of the increase would come from subalpine forests, including Douglas fir (*Pseudotsuga menziesii*). More extensive treatments in the Lower Colorado River Basin would be necessary to obtain an additional 76,170 acres annually, an 8 percent increase in water yield. About 92 percent of the total increase would be generated by treating about 20 percent of the chaparral (*Larrea tridentata*) and 33 percent of the ponderosa pine (*Pinus ponderosa*).

While information on the cost of producing extra water is incomplete, it is believed the cheapest water (based on cost to produce the additional water) would come from commercial forests, where timber yields would pay for part of the treatment costs. Water would be more expensive from vegetation conversion treatments, because most of the treatment costs would be levied against water production. Regardless, most of the water is expected to cost less than imported water and some of the water from commercial forests would supplement and be in the price range of water produced by weather modification.

In summary, modifications are needed if the current population and recreational and commercial uses are to be maintained or increased. Even with modification, there are limits. These limits must be researched carefully to preserve the natural wonders and the current uses.

Side Bar 10: Overview of transpiration

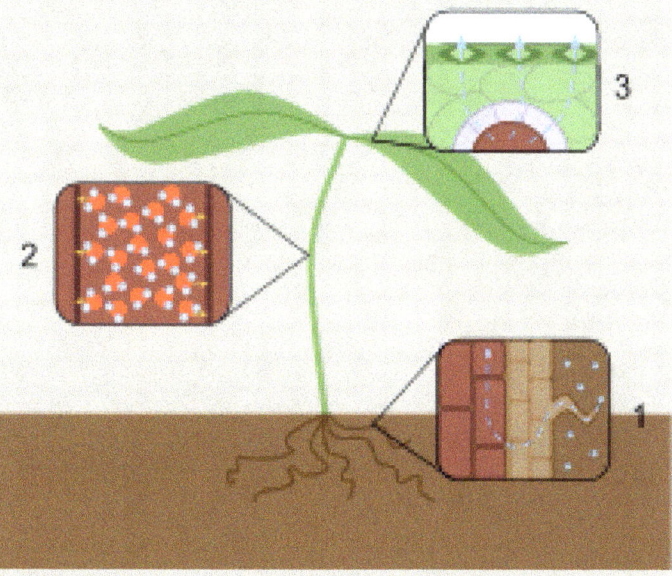

1. Water is passively transported into the roots and then into the **xylem**.

2. The forces of cohesion and adhesion cause the water molecules to form a column in the xylem.

3. Water moves from the xylem into the mesophyll cells, evaporates from their surfaces and leaves the plant by diffusion through the stomata

Glossary

American Cordillera – is a chain of mountain ranges (cordilleras) that consists of an almost continuous sequence of mountain ranges that form the western "backbone" of North America, South America and West Antarctica. It is also the backbone of the volcanic arc that forms the eastern half of the Pacific Ring of Fire.

Chert – is a fine-grained sedimentary rock composed of microcrystalline or cryptocrystalline silica. Depending on its origin, it can contain either microfossils, small macrofossils, or both. It varies greatly in color (from white to black), but most often manifests as gray, brown, grayish brown and light green to rusty red; its color is an expression of trace elements present in the rock, and both red and green are most often related to traces of iron (in its **oxidized** and reduced forms, respectively).

Clearcutting – a forestry/logging practice in which all trees in an area are uniformly cut down.

Confluence – the flowing together of two or more streams.

Disconformity – the surface of a division between parallel rock strata, indicating interruption of sedimentation: a type of unconformity

Dolomite – the name of a rock known as dolomite, dolostone, or dolomite rock.

Erosion – the action of surface processes that removes soil, rock , or dissolved material from one location on the Earth's crust, and then transport it away to another location (not to be confused with weathering which involves no movement).

Estuary (estuaries) – is a partially enclosed coastal body of water where one or more rivers or streams flow into ocean or salt water, with a free connection to the open sea

Geomorphic processes–The physical and chemical interactions between the Earth's surface and the natural forces acting upon it to produce landforms. The processes are determined by such natural environmental variables as geology, climate, vegetation and base level, to say nothing of human interference. The nature of the process and the rate at which it operates will be influenced by a change in any of these variables.

Geomorphology – the scientific study of the origin and evolution of topographic and bathymetric features created by physical, chemical or biological processes operating at or near the Earth's surface.

Graben – a depressed block of the Earth's crust bordered by parallel faults.

Gradient – upgrade, slant or slope of a hill or land.

Gypsum – a soft sulfate mineral widely mined and is used as a fertilizer, and as the main constituent in many forms of plaster, blackboard chalk and wallboard.

Homogeneity and heterogeneity – concepts often used in the sciences and statistics relating to the uniformity in a substance or organism.

Horst – A horst is a raised block of the Earth's crust that has lifted, or has remained stationary, while the land on either side (**graben**) has subsided

Karst – is a topography formed from the dissolution of soluble rocks such as limestone, dolomite, and gypsum.

Lens – a body of ore or rock or a deposit that is thick in the middle and thin at the edges.

Limestone – a sedimentary rock, composed mainly of skeletal fragments of marine organisms such as coral, forams and molluscs.

Oxidize – to add oxygen to a metal.

Pluvial – either a modern climate characterized by relatively high precipitation, or an interval of time of variable length – decades to thousands of years – during which a climate is characterized by either relatively high precipitation or humidity.

Precambrian – an informal name for the vast expanse of time prior to the Phanerozoic Eon, which includes the Paleozoic, Mesozoic, and Cenozoic Eras.

Proterozoic Eon – The period of Earth's history that began 2.5 billion years ago and ended 542.0 million years ago is known as the Proterozoic, which is subdivided into three eras: the Paleoproterozoic (2.5 to 1.6 billion years ago), Mesoproterozoic (1.6 to 1 billion years ago), and Neoproterozoic (1 billion to 542.0 million years ago).

Regolith – a layer of loose, heterogeneous superficial deposits covering solid rock. It includes dust, soil, broken rock, and other related materials and is present on Earth, the Moon, Mars, some asteroids, and other terrestrial planets and moons.

Sedimentary rocks – types of rock that are formed by the deposition and subsequent cementation of that material at the Earth's surface and within bodies of water. Sedimentation is the collective name for processes that cause mineral or organic particles (detritus) to settle in place. The particles that form a sedimentary rock by accumulating are called sediment. Before being deposited, the sediment was formed by weathering and erosion from the source area, and then transported to the place of deposition by water, wind, ice, mass movement or glaciers, which are called agents of denudation. Sedimentation may also occur as minerals precipitate from water solution or shells of aquatic creatures settle out of suspension.

Topography – the detailed mapping or charting of the features of a relatively small area, district, or locality.

Unconformity – is a buried erosional or non-depositional surface separating two rock masses or strata of different ages, indicating that sediment deposition was not continuous.

Xylem – a compound tissue in vascular plants that helps provide support and that conducts water and nutrients upward from the roots, consisting of tracheids, vessels, parenchyma cells, and woody fibers.

Side Bars

Side Bar 1   Major events in a brief geological time scale

Side Bar 2   *Geologic Time Scale Explained*

Side Bar 3   Vulcan's Throne volcano above Lava Falls

Side Bar 4   The rock record at Grand Canyon

Side Bar 5   Time Periods by Name

Side Bar 6   The "Crossing of the Fathers", now lies under Lake Powell

Side Bar 7   Edward Fitzgerald "Ned" Beale

Side Bar 8   The Central Arizona Project

Side Bar 9   Recommendations/Suggestions for water in the Grand Canyon National Park

Side Bar 10   Overview of transpiration

List of Illustrations

Fig. 1 Unconformity

Fig. 2 Shallow subduction – Laramide orogeny

Fig. 3 Plates of the crust of Earth

Fig. 4 Internal Structure of the Earth

Fig. 5. The Colorado River had cut down to nearly the current depth of the Grand Canyon by 1.2 million years ago

Fig. 6. Uplift of the Colorado Plateaus

Fig. 7. Gulf of California

Fig. 8. Map of Colorado River between Lake Powell and Lake Mead

Fig. 9. Sediment formed the Grand Canyon

Fig. 10. Bright Angel Canyon

Fig. 11. Sub-divisions of the Colorado Plateau

Fig. 12. Canyon Walls

Fig. 13. Temple Butte Limestone

Fig. 14. Redwall Limestone

Fig. 15. Surprise Canyon Formation

Fig. 16a, b. The Supai Group

Fig. 17. Animal tracks

Fig. 18. Fern fossil

Fig. 19. Four Corners of the Colorado Plateau

Fig. 20. Wagon road along the 35th parallel

Fig. 21. Rapids in the Powell expedition

Fig. 22. Marble Canyon

Fig. 23. Canyon Walls

Fig. 24. Railroad from Williams, Arizona to the South Rim

Fig. 25. Glen Canyon Dam

Fig. 26. Glen Canyon and Grand Canyon Map

Fig. 27a. Grand Canyon Uranium Mine

Fig. 27b. Uranium Mining-Havasupai Tribe

Fig. 28. White Mesa Uranium Mill

Fig. 29. Effects of grazing and browsing in trees

Fig. 30. Grand Canyon National Park is a patchwork of different management types

Fig. 31. Bison on the North Rim

Fig. 32. Viewing spots on the South Rim

Fig. 33. Coal-fired Power Plant in Page, AZ

Fig. 34. The Colorado River Basin in Grand Canyon, Arizona

Fig. 35. Clearcutting

Illustration Credits

Anthony, Alex. Northern Arizona University, "Crossing of the Fathers," *Intermountain Histories*, accessed February 1, 2019, https://www.intermountainhistories.org/items/show/65

Geologic Time Scale (Side Bar 1)- Allen G. Collins created this page, 11/26/94; Robert Guralnick and Brian R. Speer made revisions, 9/15/95; Brian R. Speer made further modifications, 6/4/98; Allen G. Collins reordered the time units with younger times above older times, 12/14/98; Sarah Rieboldt updated the page using the Geological Society of America (GSA) 1999 Geologic Timescale, 11/2002; Dave Smith created a new geologic time table using the ICS dates, adapted the page to the new site format, and made some content changes, 5/26/2011

https://3dparks.wr.usgs.gov/

https://www.grandcanyontrust.org/grand-canyon-uranium

https://en.wikipedia.org/wiki/Grand_Canyon

https://en.wikipedia.org/wiki/Central_Arizona_Project

https://en.wikipedia.org/wiki/Horst_(geology)

https://www.nps.gov/grca/learn/nature/fossils.htm

https://www.sltrib.com/news/environment/2018/10/21/ute-tribal-members-

https://www.azcentral.com/story/news/local/arizona-environment/2018/11/26/uranium-mines-near-grand-canyon-hazardous-wildlife/1633529002/

https://watershed.ucdavis.edu/education/classes/files/content/page/A%20history%20of%20grazing%20in%20and%20around%20Grand%20Canyon%20National%20Park.pdf

https://pubs.usgs.gov/wri/wri024080/pdf/WRIR4080.pdf

https://www.nps.gov/grca/learn/nature/**bison**.htm

https://www.education.vic.gov.au/school/teachers/teachingresources/discipline/science/conti nuum/Pages/geological.aspx

http://www.**old**earth.org/.../earth_history_c13_neogene_**grand_canyon_2**.htm

http://archive.library.nau.edu/cdm/ref/collection/cpa/id/62520 Northern Arizona University Cline Library Richard W. Sprang Collection.

https://en.wikipedia.org/wiki/Edward_Fitzgerald_Beale

Roberts, Matthew, Charlie Schlinger and Steve Mead. 2015. An Investigation of Energy, Use, Potable Water and Wastewater Treatment at Grand Canyon National Park, Arizona. http://www.waterenergy.nau.edu/docs/grandCanyon.pdf

Sources

Abruzzi, W.S. 1995. The Social and Ecological Consequences of Early Cattle Ranching in the Little-Colorado River Basin. Human Ecology 23:75–98.

Baker Jr., Malchus B. and Peter F. Ffolliott. 2000. Contributions of Watershed Management Research to Ecosystem-Based Management in the Colorado River Basin *(PDF). USDA Forest Service Proceedings RMRS–P–13. U.S. Forest Service, Washington, DC.* www.fs.fed.us/rm/pubs/rmrs_p013/rmrs_p013_117_128.pdf

Benke, Arthur C. and Colbert E. Cushing. 2005. Rivers of North America. Academic Press, New York, NY.

Beus, Stanley S. and Michael Morales. 2002. Grand Canyon Geology, Second Edition. Oxford University Press, New York, NY

"Boundary Descriptions and Names of Regions, Subregions, Accounting Units and Cataloging Units". *U.S. Geological Survey.* https://water.usgs.gov/GIS/huc_name.html

Charles, Grace. 2016. https://watershed.ucdavis.edu/.../ A%20history%20of%20**grazing**%20in%20and%20around%20**Gran**...

Chronic, Halka. 2004. Pages of Stone: Geology of the Grand Canyon and Plateau Country National Parks and Monuments (2nd ed.). The Mountaineers Books, Seattle, WA.

"Executive Summary, Green River Basin Water Plan". Wyoming State Water Plan. February 2001.

"Grand Canyon National Park, Arizona: Geology Fieldnotes". 2013. U.S. National Park Service.

Gupta, Avijit. 2007. Large Rivers: Geomorphology and Management. John Wiley and Sons, New York, NY.

http://www.nature.nps.gov/geology/parks/grca/

https://en.wikipedia.org/wiki/Grand_Canyon

https://www.grandcanyontrust.org/grand-canyon-uranium

http://www.history.com/topics/grand-canyon

https://www.nature.nps.gov/geology/parks/grca/

www.nps.gov

http://www.bobspixels.com/kaibab.org/geology/gc_geol.htm

http://www.eionet.europa.eu/gemet/en/concept/3652

https://en.wikipedia.org/wiki/Sedimentary_rock

https://en.wikipedia.org/wiki/Pluvial

http://www.ucmp.berkeley.edu/precambrian/proterozoic.php

http://www.ucmp.berkeley.edu/help/timeform.php

http://www.unco.edu/hewit/DOHIST/puebloan/begin.htm

https://www.revolvy.com/page/Pecos-Classification

https://pubs.usgs.gov/wri/wri024080/pdf/WRIR4080.pdf

https://web.archive.org/web/20111125065800/http:/waterplan.state.wy.us/plan/green/execsumm.html

https://web.archive.org/web/20111125065800/ http://waterplan.state.wy.us/plan/green/execsumm.html

Hibbert, A. R. 1979. Managing vegetation to increase flow in the Colorado River Basin. USDA Forest Service, General Technical Report RM-66.

Hickman, K.R., D.C. Hartnett, R.C. Cochran and C.E. Owensby. 2004. Grazing management effects on plant species diversity in tallgrass prairie. Journal of Range Management 57:58–65.

Isacks, Bryan L., Richard A. Kissel and Warren D. Allmon. 2016. Chapter 4: Topography of the Southwestern US. Pages 159-200 *In*: Swaby, A. N., M. D. Lucas, and R. M. Ross (editors). The Teacher-Friendly Guide to the Earth Science of the Southwestern US. Paleontological Research Institution (Special Publication 50), Ithaca, New York, NY

McGregor, John Charles. 1951. The Cohonina culture of northwestern Arizona. University of Illinois Press.

Price, L. Greer. 1999. Geology of the Grand Canyon. Grand Canyon, Arizona: Grand Canyon Association, Grand Canyon, AZ.

Ranney, Wayne. 2005. Carving Grand Canyon, Evidence, Theories, and Mystery. Grand Canyon Association, Grand Canyon, AZ.

Rusho, W.L. 1992. "Lee's Ferry, Arizona". Utah History Encyclopedia. University of Utah. Archived on January 9, 2013.

Stegner, Page. 1994. Grand Canyon, the Great Abyss. HarperCollins, New York, NY

Whiting, A.F. 1985. Havasupai Habitat. University of Arizona Press, Tucson, Arizona

Zandt, G., S. C. Myers and T. C. Wallace. 1995. Crust and mantle structure across the Basin and Range–Colorado Plateau boundary at 37°N latitude and implications for Cenozoic extensional mechanism. Journal of Geophysical Research 100:10529–10548

About the Author

Mary Jo Nickum is a retired librarian, teacher, writer and editor. She is also a biologist, specializing in fish and other aquatic as well as terrestrial life. She enjoys writing about biological subjects for kids. She has won Excellence in Craft awards from the Outdoor Writers Association of America (OWAA) for her books and magazine articles.

Visit her website www.asktheanimallady.com for more about animals.

www.ingramcontent.com/pod-product-compliance
Lightning Source LLC
Chambersburg PA
CBHW060426010526
44118CB00017B/2375